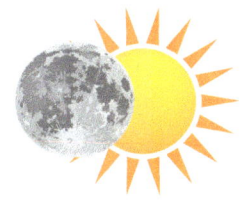

Pennsylvania
Total Eclipse Guide

Official Commemorative 2024 Keepsake Guidebook

2024 Total Eclipse State Guide Series

Large Print Edition

Aaron Linsdau

Sastrugi Press

Jackson Hole

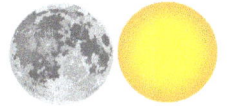

Sastrugi Press / Published by arrangement with the author
Pennsylvania Total Eclipse Guide: Official Commemorative 2024 Keepsake Guidebook

The author has made every effort to accurately describe the locations contained in this work. Travel to some locations in this book is hazardous. The publisher has no control over and does not assume any responsibility for author or third-party websites or their content describing these locations, how to travel there, nor how to do it safely. Refer to local regulations and laws.

Any person exploring these locations is personally responsible for checking local conditions prior to departure. You are responsible for your own actions and decisions. The information contained in this work is based solely on the author's research at the time of publication and may not be accurate in the future. Neither the publisher nor the author assumes any liability for anyone climbing, exploring, visiting, or traveling to the locations described in this work. Climbing is dangerous by its nature. Any person engaging in mountain climbing is responsible for learning the proper techniques. The reader assumes all risks and accepts full responsibility for injuries, including death.

Sastrugi Press
PO Box 1297, Jackson, WY 83001, United States
www.sastrugipress.com
Quantity sales: Special discounts are available on quantity purchases by corporations, associations, and others. For details, contact the publisher at the address above.
Library of Congress Catalog-in-Publication Data
Library of Congress Control Number: 2019949787
Linsdau, Aaron
Pennsylvania Total Eclipse Guide / Aaron Linsdau-1st United States edition
p. cm.
1. Nature 2. Astronomy 3. Travel 4. Photography
Summary: Learn everything you need to know about viewing, experiencing, and photographing the total eclipse in Pennsylvania on April 8, 2024.
ISBN-13: 978-1-944986-31-5 (paperback)
ISBN-13: 978-1-944986-93-3 (large print)

508.4—dc23
Printed in the United States of America when purchased in the United States
All photography, maps and artwork by the author, except as noted.
00061
10 9 8 7 6 5 4 3 2 1

Contents

Introduction

Thank you for purchasing this book. It has everything you need to know about the total eclipse in Pennsylvania on April 8, 2024.

A total eclipse passing through the United States is a rare event. The last US total eclipse was in 2017. It traveled from Oregon to South Carolina. The last American total eclipse prior to that was in 1979!

The next total eclipse over the US will not be until April 8, 2024. It will pass over Texas, the Midwest, and on to Maine. After that, the next coast-to-coast total eclipse will be in 2045.

It's imperative to make travel plans early. You will be amazed at the number of people swarming to the total eclipse path. Some might say watching a partial versus a total eclipse is a similar experience. It's not.

This book is written for Pennsylvania visitors and anyone else viewing the eclipse. You will find general planning, viewing, and photography information inside. Should you travel to the eclipse path in Pennsylvania in April, be prepared for an epic trip. The estimates based on the 2017 eclipse suggest that millions will converge on Pennsylvania.

Some hotels in the communities and cities along the path of totality in Pennsylvania have already been contacted by people to make reservations. Finding lodging along the eclipse path may be a major challenge.

Resources will be stretched far beyond the normal

limits. Think gas lines from the late 1970s. It may be likely that traffic along highways will come to a complete standstill during this event. Be prepared with backup supplies.

Many smaller Pennsylvania towns are far from any major city. Pennsylvania country roads can be slow. Please obey posted speed limits for the safety of everyone. Be cautious about believing a map application's estimate of travel time in Pennsylvania.

People in communities along the path of the total eclipse may rent out properties for this event. With this major celestial spectacle in the spring of 2024, be assured that Pennsylvania "hasn't seen anything yet."

Is this to say to avoid Pennsylvania or other areas during the eclipse? Not at all! This guidebook provides ideas for interesting, alternative, and memorable locations to see the eclipse. It will be too late to rush to a better spot once the eclipse begins. Law enforcement will be out to help drivers reconsider speeding.

Please be patient and careful. There will be a large rush of people from all over the world, converging on Pennsylvania to enjoy the total eclipse. Be mindful of other drivers on eclipse weekend, as they may not be familiar with Pennsylvania roads.

You should feel compelled to play hooky on April 8. Ask for the day off. Take your kids out of school. They'll be adults before the next chance to see a total eclipse over America. Create family memories that will last a lifetime. Sastrugi Press does not normally advocate skipping school or work. Make an exception because this is too big an event to miss.

Wherever you plan to be along the total eclipse path, leave early and remember your eclipse glasses. People from all around the planet will converge on Pennsylvania. Be good to your fellow humans and be safe. We all want to enjoy this spectacular show.

Visit www.sastrugipress.com/eclipse for the latest updates for this state eclipse book series.

Author Information

Polar explorer and motivational speaker Aaron Linsdau's first book, *Antarctic Tears*, is an emotional journey into the heart of Antarctica. He ate two sticks of butter every day to survive. Aaron coughed up blood early in the expedition and struggled with equipment failures. Despite the endless difficulties, he set a world record for surviving the longest solo expedition to the South Pole.

Aaron teaches how to build resilience to overcome adversity by managing attitude. He shares his techniques for overcoming adrenaline burnout and constant overload. He inspires audiences to face their challenges with a new perspective. As a motivational speaker, Aaron talks about courage, resilience, attitude, safety, and risk. He hopes that you will be inspired and have an enjoyable time watching the total eclipse in Pennsylvania.

Visit his websites at www.aaronlinsdau.com or www.ncexped.com

All About Pennsylvania

OVERVIEW OF PENNSYLVANIA

For any individual or family looking for an interesting total eclipse experience, Pennsylvania will be hard to beat. From the bustling cities of Philadelphia and Pittsburgh to the beautiful forests and beaches of the northeastern part of the state, there

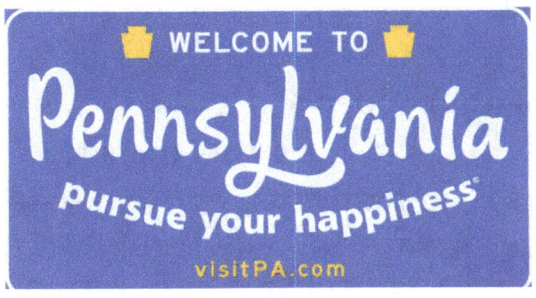

is something of interest for everyone in the Quaker State.

Pennsylvania was founded as an English colony in 1681 with a land grant from the king to William Penn, a famous Quaker, as a haven for religious tolerance. The state played a pivotal role in the American Revolution when the Founders met in Philadelphia for the First Continental Congress in 1774. They met again for the Second Continental Congress in 1775, where they wrote and signed the Declaration of Independence. Philadelphia served as the capital for the new nation until captured by the British when the Continental Congress fled to Lancaster County and then York. Several Revolutionary battles were fought in Pennsylvania, and General Washington made Valley Forge his headquarters during the winter of 1777-78.

After the war, the Articles of Confederation and

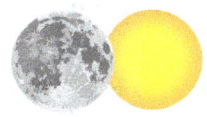

then the Constitution were written in Independence Hall in Philadelphia, the same place where the Declaration of Independence was written and signed. Pennsylvania became the second state in the new nation when the Constitution was ratified on December 12, 1787. Today tourists can tour historic Philadelphia and see Independence Hall and the Liberty Bell. Valley Forge is also a great destination for history fans, where they can learn about the hardships and struggles faced by the Continental Army in the fight for independence.

Nearly a century later, Gettysburg was the site of one of the most harrowing battles of the Civil War when the Confederate Army attempted to invade the North in July 1863. About 50,000 men on both sides died over the three-day battle, and it was the site of Abraham Lincoln's Gettysburg Address, one of the most famous speeches in American history.

For modern history buffs, Philadelphia is across the river from Camden, New Jersey, home of the USS New Jersey Museum and Memorial. Visitors can tour the historic battleship and learn about World War II and the famous ship's role in it and later conflicts into the 1980s. Also located in Philadelphia is the Philadelphia Museum of Art, home of the famous stairs from Rocky and a statue of the character Rocky Balboa from the film. Professional football (Eagles), basketball (76ers), baseball (Phillies), and hockey (Flyers) teams, as well as great dining options, make Philadelphia worth a visit.

Another Pennsylvania destination that is not to be missed is historic Hershey Park, home of the world-famous chocolate factory and amusement park. Located between Philadelphia and Pittsburgh, Hershey Park is a top-notch family destination. Nearby is Lancaster, the center of the Pennsylvania Dutch community. Visitors flock to Lancaster to enjoy the relaxing rural atmosphere and learn about the Amish community.

Lovers of nature will also enjoy exploring the many beaches along Lake Erie. The best place to experience the lake is Presque Isle State Park near Erie in the northeastern corner of the state. Also located at the park is the Tom Ridge Environmental Center, where visitors can learn about the ecosystem and history of this important location.

Pittsburgh is the second of Pennsylvania's major cities. In addition to dining options from blue collar to world class, Pittsburgh hosts the Andy Warhol Museum and the Phipps Conservatory. Phipps contains fourteen glass rooms representing the fauna of many different ecosystems around the world, along with twenty-three outdoor gardens and one of the world's best collections of orchids. No mention of Pittsburgh is complete without sports. Pittsburgh is famous for the Steelers (football), Pirates (baseball), and Penguins (hockey).

As you can see, Pennsylvania is brimming with plenty of places for individuals and families to visit. No matter what time of year, Pennsylvania has something for everyone and should not be missed.

Hotels and Motels During the Eclipse

Once excitement of the total eclipse over Pennsylvania spreads, rooms will become scarce. Many hotels in towns along the path of totality in western states sold out for a year or more during the 2017 total eclipse. Pennsylvania is not alone in this challenge. Hotels all along the path of totality will sell out in anticipation of the 2024 total eclipse.

What does this mean for eclipse visitors? Lodging and room rentals in eclipse towns will be at a massive premium. Does that mean all hope is lost to find a place to stay? Not at all. But you will have to be creative. There will be few, if any, hotel rooms available in these eclipse cities by early 2024. Accommodations in the cities and towns along the path of the eclipse will be difficult to come by.

In summer 2017, the author searched on Hotels.com for rooms along the 2017 total eclipse path on the weekend of August 21 and found many major cities sold out. Once word of the 2024 eclipse spreads, room rates will increase and availability will drop.

Search for rooms farther away from the eclipse path. If you are willing to stay in cities outside the eclipse path, you will have better success at finding rooms. As the eclipse approaches, people will book rooms farther from the totality path. By early spring, rooms in cities near the total eclipse path may be unavailable. The effect of this event will be felt across Pennsylvania and the rest of the United States.

Think regionally when looking for rooms. Be prepared to search far and wide during this major event. If a five-hour drive is manageable, your lodging options greatly expand, but it also increases your travel risk.

Internet Rentals

To find rooms to stay in towns along the eclipse path, try a web service such as Airbnb.com. Note that some people rent out rooms or homes illegally, against zoning regulations. Cities will feel the crunch of inquiries early due to others who experienced the 2017 eclipse.

If cities fully enforce zoning laws, authorities may prevent your weekend home rental. Online home rentals during the eclipse will be a target for rental scams. People from out of the area steal photos and descriptions, then post the home for rent. You send your check or wire money to a "rental agent" then show up to find you have been scammed. If the deal sounds strange or too good to be true, run away.

Camping

If you can book a campsite, do it as soon as you can. Do not wait. All areas in the national forests are first-come, first-served. Forest roads may be packed. Expect all areas to be swarming with people. Show up early to stake out your spot. Consider staying farther away and driving early on April 8.

Please respect private land too. Pennsylvania folks

don't take kindly to people overrunning their property without permission. In a big state with millions of residents, people are very protective, but they're friendly, too. You never know what you might be able to arrange with a smile and a bit of money.

This all said, there are plenty of camping opportunities throughout Pennsylvania. You don't have to sleep exactly on the eclipse path. If you're ready to rough it, there are national forest camping options.

Government agencies will meet years in advance to talk about how to manage the influx of people. Every possible government agency will be working full time to enforce the various rules and regulations.

National Parks and Monuments

Finding a camping site at any state park, national park, or national monument along the eclipse path in Pennsylvania will be challenging. To watch the eclipse from any location, you do not have to sleep in it. You just need to drive there in the morning.

Law enforcement will be present on the eclipse weekend. Hundreds of thousands of people are expected in the region. Parking may overflow. It will make parking lots and lines on Black Friday at the mall look uncrowded. For an event of this magnitude, find your location as early as possible.

The first sentence of the national parks mission statement is:

"The National Park Service preserves

unimpaired the natural and cultural resources and values of the national park system for the enjoyment, education, and inspiration of this and future generations."

Roadside camping (sleeping in your car) is not allowed in national monuments or parks. Park facilities are only designed to handle so many people per day. Water, trash collection, and toilets can only withstand so much. If you notice trash on the ground, take a moment to throw it away. Protect your national park and help out. Rangers are diligent and hardworking but they can only do so much to manage the expected crowds.

NATIONAL AND STATE FORESTS

There are national and state forest options in Pennsylvania. They all have camping opportunities. The forest service manages undeveloped and primitive campsites. Be sure to check for any fire restrictions. Check with individual agencies for last-minute information and regulations. The forest service requires proper food storage. Plan to purchase food and water before choosing your campsite. Below is a partial list of national forests in Pennsylvania:

Allegheny National Forest: https://www.fs.usda.gov/ allegheny
Susquehannock State Forest
Buchanan State Forest
Tiadaghton State Forest

Backcountry service roads abound in Pennsylvania. Maps for forests are available at local visitor centers and bookstores. This book's website has digital copies of some forest maps.

Printed national forest maps are large and detailed. They have illustrated road paths, connections, and other vital travel information not available on digital device maps. Viewing digital maps on your smartphone or mobile pad is difficult. If you plan to camp in the forest, a real paper map is a wise investment.

Camping in federal wilderness areas is also allowed. Those areas afford the ultimate backcountry experience. However, be aware that no vehicle travel is allowed in the specially designated areas. This ban includes: vehicles, bikes, hang gliders, and drones. You can travel only on foot or with pack animals.

Sleep in Your Car

Countless RVs, campers, trucks, cars, and motorcycles will flood Pennsylvania. Sleeping in your car with friends is tolerable. Doing so with unadventurous spouses or children is another matter.

Do not be caught along the path of the total eclipse without some sort of plan, especially in the bigger cities of Pennsylvania. The whole path of totality will fill with people on April 8.

Useful Local Webcams

Local webcams are handy to make last-minute travel decisions. The webcams are sensitive enough to show headlights at night. Use them to determine if there are issues before traveling out. Eclipse traffic will add to the morning commuter traffic.

The smartphone application Wunderground is useful to check on webcams in one place. Selected the webcams are listed in the app. Whether you use this app or another, an Internet search will reveal many useful webcams for your search.

Weather

It's all about the weather during the eclipse. Nothing else will matter if the sky is cloudy. You can be nearly anywhere along the path in Pennsylvania and catch a view of the event when traffic comes to a standstill. But if there's a cloud cover forecast, seriously reconsider your viewing location.

Travel early wherever you plan to go. Attempting to change locations an hour before the eclipse due to weather will likely cause you to miss the event. Pennsylvania country roads can be narrow and slow. The number of vehicles will cause unexpected back-ups.

MODERN FORECASTS

Use a smartphone application to check the up-to-date weather. Wunderground is a good application and has relatively reliable forecasts for the region. The hourly forecast for the same day has been rather accurate for the last two years. The below discussion refers to features found in the Wunderground app. However, any application with detailed weather views will improve your eclipse forecasting skills.

Cloud Cover Forecast

The most useful forecast view is the visible and infrared cloud-coverage map. Avoid downloading this app the night before and trying to learn how to read it. Practice reading them at home. It's imperative to understand how to interpret the maps early.

All cloud cover, night or day, will appear on an infrared map. Warm, low-altitude clouds are shown in white and gray. High-altitude cold clouds are displayed in shades of green, yellow, red, and purple. Anything other than a clear map spells eclipse-viewing problems.

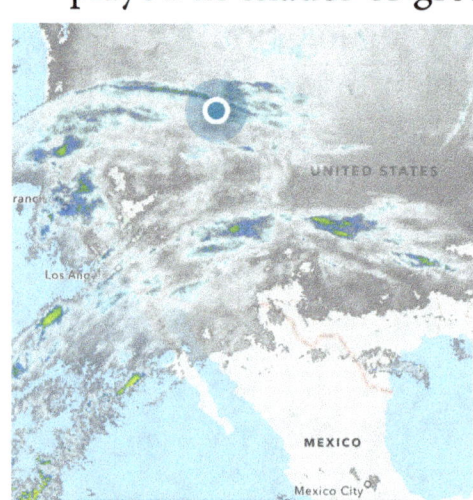

To improve your weather guess, use the animated viewer of the cloud cover. It will give you a sense of cloud motion. You can discern whether clouds or rain are moving toward, away from, or circulating around your location.

Infrared cloud map showing the worst case eclipse cloud cover. Courtesy of National Weather Service.

Normal Pennsylvania

WEATHER PATTERN

Due to the direction of the jet stream, most weather travels across the Pacific Ocean, through the western states, over the Rockies, and then into Pennsylvania. On occasion, weather can approach from Canada or Mexico. Due to the nature of the storms from the Arctic, weather in Pennsylvania can be unpredictable.

The common weather pattern in April is slightly warm in the afternoon and mildly cool in the evenings. Passing cold fronts in spring can bring unexpected cloud cover and rains.

Historically, Pennsylvania tends to have moderate cloud cover during April. Prepare to make adjustments. If anything other than clear skies are predicted, drive to other parts of Indiana, Missouri, or Pennsylvania.

Be aware of tornadoes in Pennsylvania. Although the peak tornado season is June, there have been many recorded tornadoes in April. Pay attention to the weather forecast. If dangerous weather is predicted, your main concern should be safety rather than chasing an eclipse.

Consider that slow-moving clouds can obscure the sun for far longer than the four-minute duration of the totality. The time of totality is so short that you do not want to risk it. Missing it due to a single cloud will be a major disappointment.

LOCAL ECLIPSE WEATHER FORECASTS

Local town and city newspapers, radio, and television stations around Pennsylvania will have a weekend edition with articles discussing the eclipse weather. However, conditions change unpredictably in Pennsylvania. A three-day forecast for April may be incorrect.

FINDING THE RIGHT LOCATION TO VIEW ECLIPSE EFFECTS

One of the peculiarities of total eclipses is that the entire show is not only in the sky. There are other unusual effects seen during the total eclipse that are worth looking for.

The first effect to watch for is the crescent moon shapes created from leaf shadows on the ground. They're best viewed on a sidewalk or asphalt. They can only be observed during the partial eclipse. The other effect that is worth watching for is the shadow bands or "snakes" as they're commonly called.

Shadow banding is seen right before and after the totality takes place. They're best observed on smooth, plain-colored surfaces. If you plan to be in the forest for the eclipse, you may struggle to see the bands but will likely see crescent shadows all around on the ground.

One of the supreme challenges with all of these effects is choosing what to watch. You can see the

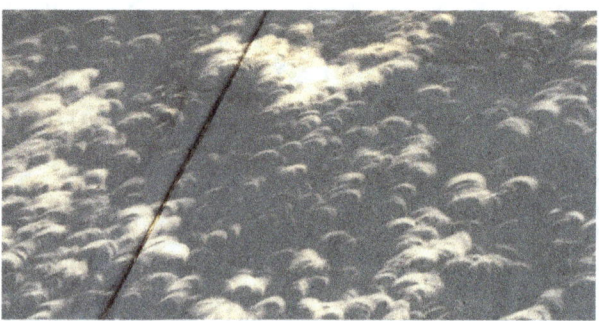

PENNSYLVANIA

crescent shadows in the hour before and after the totality but shadow banding happens before or after totality. It is more difficult to look away from the eclipse than you think.

Road Closures and Traffic

Highways connecting various Pennsylvania towns in the total eclipse path will be heavily impacted on the weekend before and day of the eclipse. As was found in the 2017 total eclipse, there is no way to predict which areas will be impacted.

Planning ahead is essential to give you the best opportunity to enjoy the eclipse without the nightmare of being stuck in traffic for hours on end. The traffic in states during the 2017 eclipse was stunning, so imagine what it will be like for Pennsylvania in 2024.

Update yourself with the latest road report information from the Pennsylvania road condition website:

https://www.511pa.com/

It's imperative to plan early and have one if not more backup plans in case of difficult travel conditions. April weather is unpredictable and variable.

If you believe it's necessary to leave a town to watch the eclipse, do so the night before or extremely early in the morning. RVs are common, and trains of them crawl through popular areas.

Communication Information

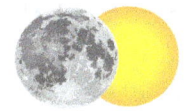

CELLULAR PHONES

Cellular "cell" phone service in some remote Pennsylvania locations may be problematic. Most of the time there is good coverage along the main highways and interstates. However, even along major thoroughfares, there can be little or no coverage.

It's possible to find zones where text messages will send when phone calls are impossible. If you cannot make a phone call, the chance of having data coverage for web surfing or e-mail is low.

Please look up any information or communicate what you need before departing from the main roads around Pennsylvania. Bureau of Land Management (BLM) areas sometimes have coverage. Planned to be self-contained. Plan for your cell phone not to connect.

You may find yourself out of cell service. With a large number of cell users in a concentrated area, coverage and data speed may collapse as well. Search on the phrase "cell phone coverage breathing".

Wilderness and Forest Safety

All Pennsylvania forest and wilderness areas are full of wild animals. Although beautiful, wild animals can be dangerous. They can easily injure or kill people, as they are far more powerful than humans. Do not try to feed any wild animals, including squirrels, foxes, and chipmunks, as they can carry diseases. These suggestions apply to all public lands.

TICKS

Ticks exist all across the United States, but not all species transmit disease. Ticks cannot fly or jump, but they climb grasses in shrubs in order to attach to people or animals that pass by. Ticks feed on the blood of their host. In doing so, they can transmit potentially life-threatening diseases such as Lyme disease.

SPIDERS

Although the chance of encountering a venomous spider is low, it is not uncommon to encounter them. The two dangerous spiders in the state are the black widow and brown recluse. Should you encounter either species, simply leave it alone. If you are bitten, seek immediate medical attention, as their toxin can be life-threatening.

VENOMOUS SNAKES

There are multiple species of venomous snakes in Pennsylvania including the Copperhead, Eastern Massasauga Rattlesnake, and the Timber (Canebrake) Rattlesnake. Although these reptiles are not generally aggressive, they can strike when provoked or threatened. Of the approximately 8,000 people annually bitten by venomous snakes in the United States, ten to fifteen people die according to the U.S. Food and Drug Administration.

The best way to avoid rattlesnake encounters is to be mindful of your environment. Do not place your hands or feet in locations where you cannot clearly see the surroundings. Avoid heavy brush or tall weeds where snakes hide during the day. Step on a log or rock rather than over it, as a hidden snake might be on the other side. Rattlesnakes may not make any noise before striking.

Avoid handling all snakes. Should you be bitten, stay calm and call 911 or emergency dispatch as soon as possible. Transport the victim to the nearest medical facility immediately. Rapid professional treatment is the best way to manage rattlesnake bites. Refer to US Forest Service and professional medical texts for more information on managing rattlesnakes injuries.

BEARS

The forests of Pennsylvania are potentially home to black bears. Though they are listed rarely seen, they have been sighted in the state. Although they often appear docile, they can become aggressive if threatened. In the unlikely event of an attack, fight back against the bear. Use whatever you have at your disposal to defend yourself. Report all negative or aggressive bears to the local authorities.

If a bear hears you, it will usually vacate the area. Bear charges are often caused by unexpected and surprise encounters. Noise is the best defense to avoid surprising bears. Regularly clap, make noise, and talk loudly. The Pennsylvania Game Comission is a good starting point to learn more about black bears at https://www.pgc.pa.gov/Wildlife/WildlifeS-

pecies/BlackBear/Pages/LivingwithBlackBears.aspx.
It is recommended to stay one hundred yards (300 feet) away from all bears. They are exciting to see but need their space. Refer to current forest or park regulations for more safety information.

Mountain Lions

Though declared to be extinct in the state, there have been news reports of mountain lions by various news agencies. If you encounter a mountain lion, do not run. Keep calm, back away slowly, and maintain eye contact. Do all you can to appear larger. Stand upright, raise your arms, or hoist your jacket. Never bend over or crouch down. If attacked, fight back.

Eclipse Day Safety

1. Hydrate

Spring temperatures are usually mild to warm. The excitement of the event can distract you from managing hydration. Drink plenty of water. Consume more than you would at home.

2. Eye Safety time

Use certified eclipse safety glasses at all times when viewing the partial eclipse. Only remove the glasses when the totality happens. Give your eyes time to rest. They can dry out and become irritated. Bring FDA approved eye drops to keep your eyes moist.

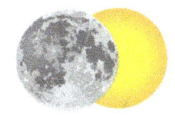

PENNSYLVANIA

3. Sun exposure

Facing at the sun for three hours can result in sunburns. Wear sunglasses and liberally apply sunscreen to avoid sunburns.

4. Eat well

Keep your energy up. Appetite loss is common when traveling. Maintain your normal eating schedule.

5. Prepare for temperature changes

Temperatures will drop rapidly during the eclipse and also once the sun sets. Bring appropriate clothing.

6. Talk with your doctor

If the humidity or heat bothers you talk with your doctor before traveling. Seek professional medical attention for serious symptoms

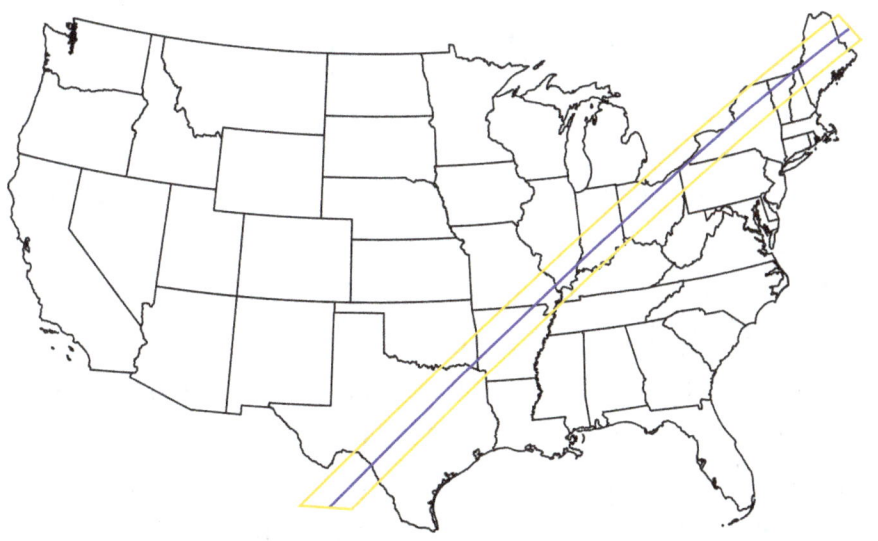

Total eclipse path across the United States (approximate).

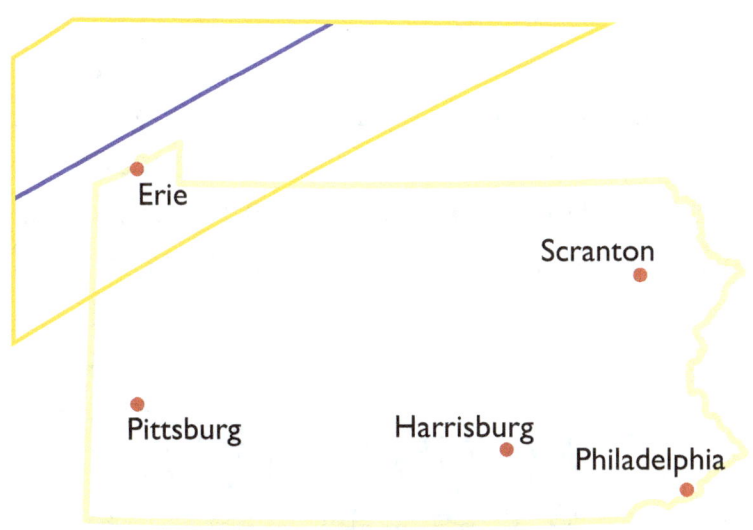

Total eclipse path across Pennsylvania (approximate).

All About Eclipses

How an Eclipse Happens

An eclipse occurs when one celestial body falls in line with another, thus obscuring the sun from view. This occurs much more often than you'd think, considering how many bodies there are in the solar system. For instance, there are over 150 moons in the solar system. On Earth, we have two primary celestial bodies: the sun and the moon. The entire solar system is constantly in motion, with planets orbiting the sun and moons orbiting the planets. These celestial bodies often come into alignment. When these alignments cause the sun to be blocked, it is called an eclipse.

For an eclipse to occur, the sun, Earth, and moon must be in alignment. There are two types of eclipses: solar and lunar. A solar eclipse occurs when the moon obscures the sun. A lunar eclipse occurs when the moon passes through Earth's shadow. Solar

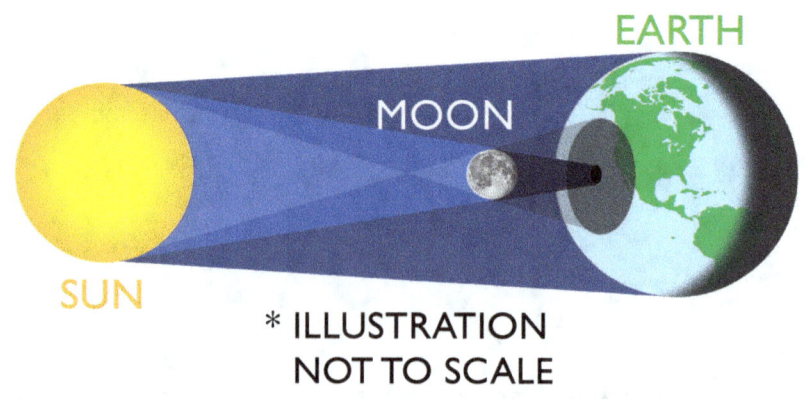

EARTH

MOON

SUN

* ILLUSTRATION
NOT TO SCALE

ECLIPSES

eclipses are much more common, as we experience an average of 240 solar eclipses a century compared to an average of 150 lunar eclipses. Despite this, we are more likely to see a lunar eclipse than a solar eclipse. This is due to the visibility of each.

For a solar eclipse to be visible, you have to be in the moon's shadow. The problem with viewing a total eclipse is that the moon casts a small shadow over the world at any given time. You have to be in a precise location to view a total eclipse. The issue that arises is that most of these locations are inaccessible to most people. Though many would like to see a total solar eclipse, most aren't about to set sail for the middle of the Pacific Ocean. In fact, a solar eclipse is visible in the same place on the world on average every 375 years. This means that if you miss a solar eclipse above your hometown, you're not going to see another one unless you travel or move.

It's much easier to catch a glimpse of a lunar eclipse, even though they occur at a much lower frequency than their solar counterparts. A lunar eclipse darkens the moon for a few hours. This is different than a new moon when it faces away from the sun. During these eclipses, the moon fades and becomes nearly invisible.

Another result of a lunar eclipse is a blood moon. Earth's atmosphere bends a small amount of sunlight onto the moon turning it orange-red. The blood moon is caused by the dawn or dusk light being refracted onto the moon during an eclipse.

ECLIPSES

Lunar eclipses are much easier to see. Even when the moon is in the shadow of Earth, it's still visible throughout the world because of how much smaller it is than Earth.

TOTAL VS. PARTIAL ECLIPSE

What is the difference between a partial and total eclipse? A total eclipse of either the sun or the moon will occur only when the sun, Earth, and the moon are aligned in a perfectly straight line. This ensures that either the sun or the moon is partially or completely obscured.

In contrast, a partial eclipse occurs when the alignment of the three celestial bodies is not in a perfectly straight line. These types of eclipses usually result in only a part of either the sun or the moon being obscured. This is often what led to ancient civilizations believing that some form of magical beast or deity was eating the sun or the moon. It appears as though something has taken a bite out of either the sun or the moon during a partial eclipse.

Total eclipses, rarer than partial eclipses, still occur quite often. It's more difficult for people to be in a position to experience such an event firsthand. Total solar eclipses can only be viewed from a small portion of the world that falls into the darkest part of the moon's shadow. Often this happens in the middle of the ocean.

THE MOON'S SHADOW

The moon's shadow is divided into two parts: the umbra and the penumbra. The former is much smaller than the latter, as the umbra is the innermost and darkest part of the shadow. The umbra is thus the central point of the moon's shadow, meaning that it is extremely small in comparison to the entire shadow. For a total solar eclipse to be visible, you need to be directly beneath the umbra of the moon's shadow. This is because that is the only point at which the moon completely blocks the view of the sun.

In contrast, the penumbra is the region of the moon's shadow in which only a portion of the light cast by the sun is obscured. When standing in the penumbra, you are viewing the eclipse at an angle. In the penumbra, the moon does not completely

ECLIPSES

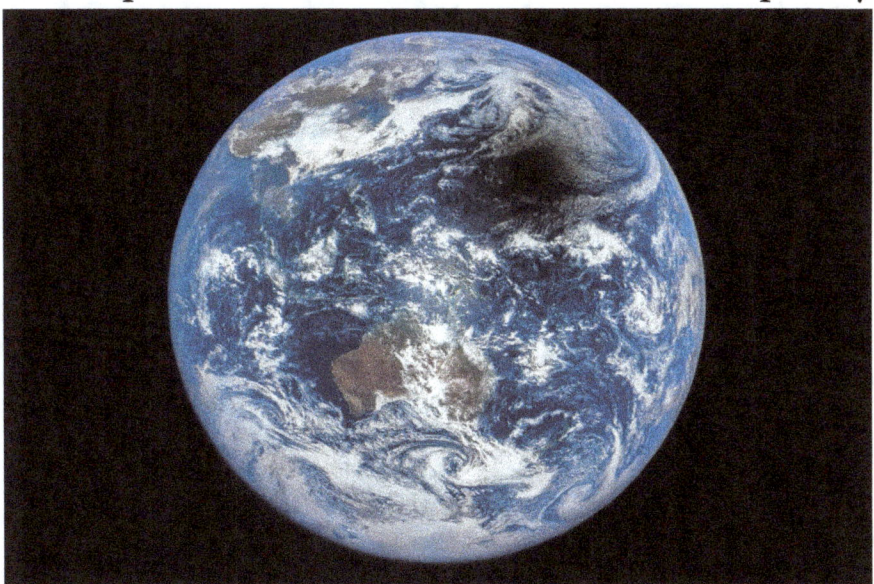

Total eclipse shadow 2016 as seen from 1 million miles on the Deep Space Climate Observatory satellite. Courtesy of NASA.

ECLIPSES

block the sun from view. This means that while the event is a total solar eclipse, you'll only see a partial eclipse. The umbra for the April 8 eclipse is over one hundred miles wide. The penumbra will cover much of the United States, Canada, and Mexico.

To provide some context, one previous total solar eclipse we experienced occurred on March 9, 2016, and was visible as a partial eclipse across most of the Pacific Ocean, parts of Asia, and Australia. However, the only place in the world to view this total solar eclipse was in a few parts of Indonesia.

Due to the varied locations and the brief periods for which they're visible, it's difficult to see each and every eclipse that occurs. Many people don't even realize that they have occurred. Consider that the umbra of the moon represents such a small fraction of the entire shadow and the majority of our planet is comprised of water. Thus, the rarity of being able to view a total solar eclipse increases significantly because it's likely that the umbra will fall over some part of the ocean rather than a populated landmass.

ECLIPSES THROUGHOUT HISTORY

Ancient peoples believed eclipses were from the wrath of angry gods, portents of doom and misfortune, or wars between celestial beings. Eclipses have played many roles in cultures, creating myths since the dawn of time. Both solar and lunar eclipses affected societies worldwide. Inspiring fear, curiosity, and the creation of legends, eclipses have cast a long shadow in the collective unconscious of humanity throughout history.

EARLY MYTH & ASTRONOMY

Documented observations of solar eclipses have been found as far back in history as ancient Egyptian and Chinese records. Timekeeping was important to ancient Chinese cultures. Astronomical observations were an integral factor in the Chinese calendar. The first observation of a solar eclipse is found in Chinese records from over 4,000 years ago. Evidence suggests that ancient Egyptian observations may predate those archaic writings.

Many ancient societies, including Roman, Greek and Chinese civilizations, were able to infer and foresee solar eclipses from astronomical data. The sudden and unpredictable nature of solar eclipses had a stressful and intimidating effect on many societies that lacked the scientific insight to accurately predict astronomical events. Relying on the sun for their agricultural livelihood, those societies interpreted solar eclipses as world-threatening disasters.

In ancient Vietnam, solar eclipses were explained as a giant frog eating the sun. The peasantry of ancient Greece believed that an eclipse was the sign of a furious godhead, presenting an omen of wrathful retribution in the form of natural disasters. Other cultures were less speculative in their investigations. The Chinese Song Dynasty scientist Shen Kuo proved the spherical nature of the Earth and heavenly bodies through scientific insight gained by the study of eclipses.

The Eclipse in Native American Mythology

Eclipses have played a significant role in the history of the United States. Before Europeans settled in the Americas, solar eclipses were important astronomical events to Native American cultures. In most native cultures, an eclipse was a particularly bad omen. Both the sun and the moon were regarded as sacred. Viewing an eclipse, or even being outside for the duration of the event, was considered highly taboo by the Navajo culture. During an eclipse, men and women would simply avert their eyes from the sky, acting as though it was not happening.

The Choctaw people had a unique story to explain solar eclipses. Considering the event as the mischievous actions of a black squirrel and its attempt to eat the sun, the Choctaw people would do their best to scare away the cosmic squirrel by making as much noise as possible until the end of the event, at which point cognitive bias would cause them to believe they'd once again averted disaster on an interplanetary scale.

Contemporary American Solar Phenomena

The investigation of solar phenomena in twentieth-century American history had a similarly profound effect on the people of the United States. A total solar eclipse occurring on the sixteenth of June,

1806, engulfed the entire country. It started near modern-day Arizona. It passed across the Midwest, over Ohio, Pennsylvania, New York, Massachusetts, and Connecticut. The 1806 total eclipse was notable for being one of the first publicly advertised solar events. The public was informed beforehand of the astronomical curiosity through a pamphlet written by Andrew Newell entitled *Darkness at Noon, or the Great Solar Eclipse.*

This pamphlet described local circumstances and went into great detail explaining the true nature of the phenomenon, dispelling myth and superstition, and even giving questionable advice on the best methods of viewing the sun during the event. Replete with a short historical record of eclipses through the ages, the *Darkness at Noon* pamphlet is one of the first examples of an attempt to capitalize on the mysterious nature of solar eclipses.

Another notable American solar eclipse occurred on June 8, 1918. Passing over the United States from Washington to Florida, the eclipse was accurately predicted by the U.S. Naval Observatory and heavily documented in the newspapers of the day. Howard Russell Butler, painter and founder of the American Fine Arts Society, painted the eclipse from the U.S. Naval Observatory, immortalizing the event in *The Oregon Eclipse.*

Four more total solar eclipses occurred over the United States in the years 1923, 1925, 1932, and 1954, with another occurring in 1959. The October

2, 1959, solar eclipse began over Boston, Massachusetts. It was a sunrise event that was unviewable from the ground level. Eminent astronomer Jay Pasachoff attributed this event to sparking his interest in the study of astronomy. Studying under Professor Donald Menzel of Williams College, Pasachoff was able to view the event from an airline hired by his professor.

To this day, many myths surround the eclipse. In India, some local customs require fasting. In eastern Africa, eclipses are seen as a danger to pregnant women and young children. Despite the mystery and legend associated with unique and rare astronomical events, eclipses continue to be awe-inspiring. Even in the modern day, eclipses draw out reverential respect for the inexorable passing of celestial bodies. They are a reminder of the intimate relationship between the denizens of Earth and the universe at large.

FUTURE ECLIPSES

The year 2017 brought the world's most-watched total eclipse in history on August 21, when a total solar eclipse crossed the United States. An annular eclipse, a "ring of fire," will pass over the United States in 2023 from Oregon to Texas. Though impressive, it will not compare to the 2024 total eclipse. There is little in nature that equals the spectacle of the sun's corona and seeing stars in the day.

There will be multiple partial, annular, or hybrid

eclipses across the world before the 2024 total eclipse. However many are in remote, inaccessible, or potentially dangerous locations on the globe. In 2019 and 2020, Chile and Argentina will experience total eclipses. The next total eclipse after that will occur over Antarctica in 2021. An extremely rare hybrid eclipse will happen in 2023 over the Indian Ocean, Australia, and Indonesia.

The next total solar eclipse viewable from the United States will occur on April 8, 2024. It will be visible in fifteen states: Texas, Oklahoma, Arkansas, Missouri, Tennessee, Kentucky, Illinois, Indiana, Ohio, Pennsylvania, Michigan, New York, Vermont, New Hampshire, and Maine.

ECLIPSES

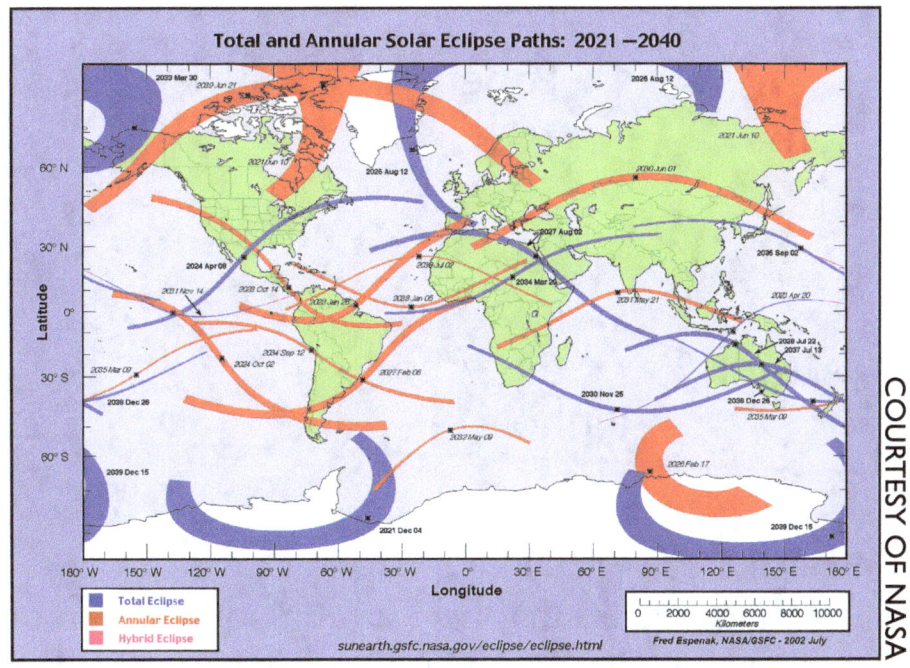

COURTESY OF NASA

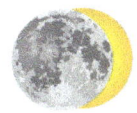

Viewing and Photographing the Eclipse

At-home Pinhole Method

Use the pinhole method to view the eclipse safely. It costs little but is the safest technique there is. Take a stiff piece of single-layer cardboard and punch a clean pinhole. Let the sun shine through the pinhole onto another piece of cardboard. That's it!

Never look at the sun through the pinhole. Your back should be toward the sun to protect your eyes. To brighten the image, simply move the back piece of cardboard closer to the pinhole. To see it larger, move the back cardboard farther away. Do not make the pinhole larger. It will only distort the crescent sun.

PHOTOGRAPHY

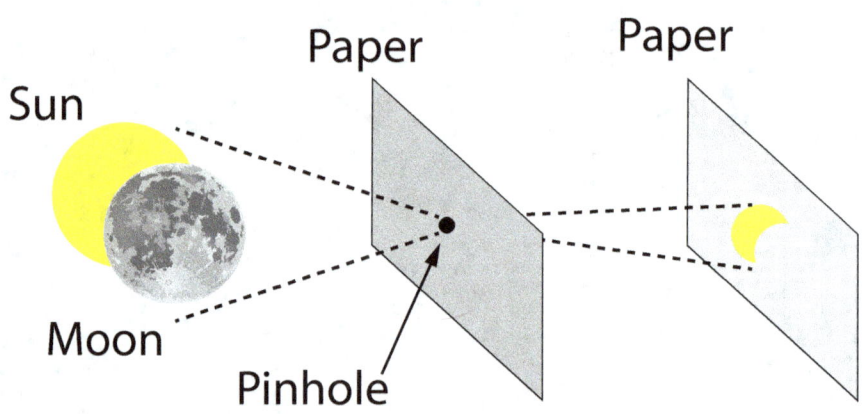

Welding Goggles

Welding goggles that have a rating of fourteen or higher are another useful eclipse viewing tool. The goggles can be used to view the solar eclipse directly.

Do not use the goggles to look through binoculars or telescopes, as the goggles could potentially shatter due to intense direct heat. Avoid long periods of gazing with the goggles. Look away every so often. Give your eyes a break.

SOLAR FILTERS FOR TELESCOPES

The ONLY safe way to view solar eclipses using telescopes or binoculars is to use solar filters. The filters are coated with metal to diminish the full intensity of the sun. Although the filters can be expensive, it is better to purchase a quality filter rather than an inexpensive one that could shatter or melt from the heat.

The filters attach to the front of the telescope for easy viewing. Remember to give your telescope cooling breaks. Rapid heating can damage your equipment with or without filters attached.

WATCH OUT FOR UNSAFE FILTERS

There are several myths surrounding solar filters for eclipse viewing. In order for filters to be safe, they must be specially designed for looking at a solar eclipse. The following are all unsafe for eclipse viewing and can lead to retinal damage: developed colored or chromogenic film, black-and-white negatives such as X-rays, CDs with aluminum, smoked glass, floppy disk covers, black-and-white film with no silver, sunglasses, or polarizing films.

PHOTOGRAPHY

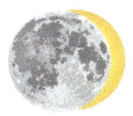

PHOTOGRAPHY

WATCH OUT FOR UNSAFE ECLIPSE GLASSES

During the 2017 total eclipse, several vendors sold eclipse glasses that were not safe for viewing the sun. Although they were marketed as safe and were even marked with the ISO 12312-2 certification, they did not block eye-damaging visible, infrared, and ultraviolet light. Check the American Astronomical Society's website (eclipse.aas.org) for a list of reputable eclipse glasses vendors.

VIEWING WITH BINOCULARS

When viewing the eclipse with binoculars, it is important to use solar filters on both lenses until totality. Only then is it safe to remove the filter. As the sun becomes visible after totality, replace the filters for safe viewing. Protect your pupils. Remember to give your binoculars a cool-down break between viewings. They can overheat rapidly from being pointed directly at the sun even with filters attached.

PLANNING AHEAD

There are many things to keep in mind when viewing a total eclipse. It is important to plan ahead to get the most out of this extraordinary experience.

UNDERSTANDING SUN POSITION

All compass bearings in this book are true north. All compasses point to Earth's magnetic north. The difference between these two measurements is called magnetic declination. The magnetic declination for

Pennsylvania is:

11° 3' W ± 0° 22' (for Harrisburg in 2024)

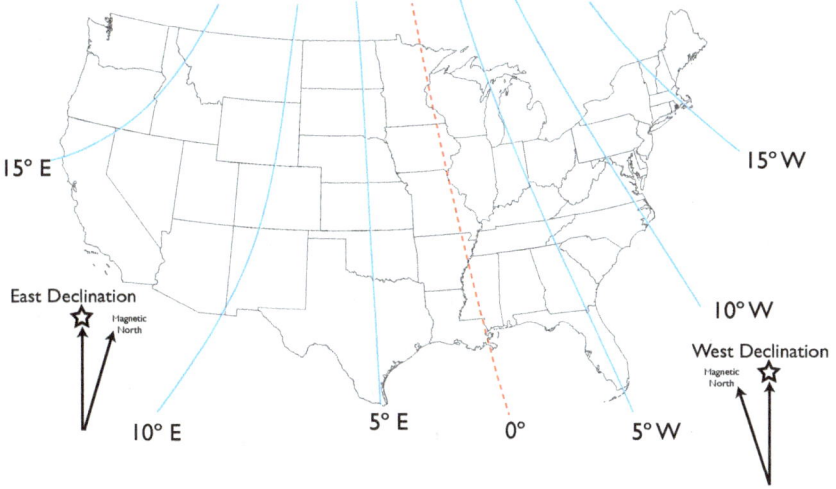

Adjust the declination from the azimuth bearing as given in the text, and set your compass to that direction.

If you purchase a compass with a built-in declination adjustment, you can change the setting once and eliminate the calculations. The Suunto M-3G compass has this correction. A compass with a sighting mirror or wire will help you make a more accurate azimuth sighting.

The Suunto M-3G also has an inclinometer. This allows you to measure the elevation of any object above the horizon. Use this to figure out how high the sun will be above your position.

You can also use a smartphone inclinometer and compass for this purpose. Make sure to calibrate your smartphone's compass before every use, oth-

PHOTOGRAPHY

erwise it might indicate the wrong bearing. Set the smartphone compass for true north to match the book. Understand the compass prior to April 8. There will be little time to guess or search on Google. Smartphone and GPS compasses are "sticky." Their compasses don't swing as freely as a magnetic compass does.

The author has used his magnetic compass for azimuth measurements and a smartphone to measure elevation. Combining these two tools will allow you to make the best sightings possible.

Outdoor sporting goods stores in most towns and cities carry compasses. Purchase and practice with a good compass in your hometown well before the event. Take the time to learn how to use it before the day of the eclipse. You do not want to struggle with orienteering basics under pressure.

Sun Azimuth

Azimuth is the compass angle along the horizon, with 0° corresponding to north, and increasing in a clockwise direction. 90° is east, 180° is south, and 270° is west.

Sun Elevation

Altitude is the sun's angle up from the horizon. A 0° altitude means exactly on the horizon and 90° means "straight up."

Using the sun azimuth and elevation data, you can predict the position of the sun at any given time.

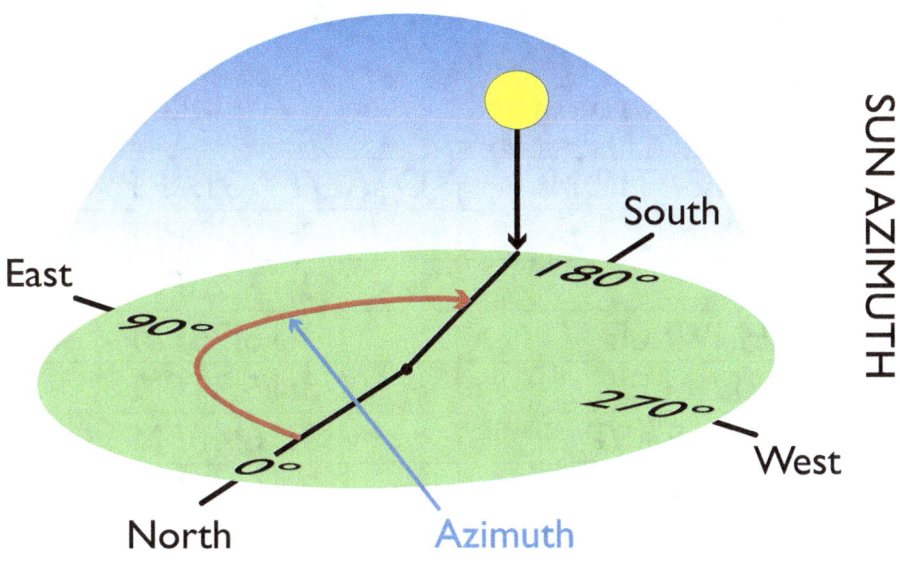

SUN AZIMUTH

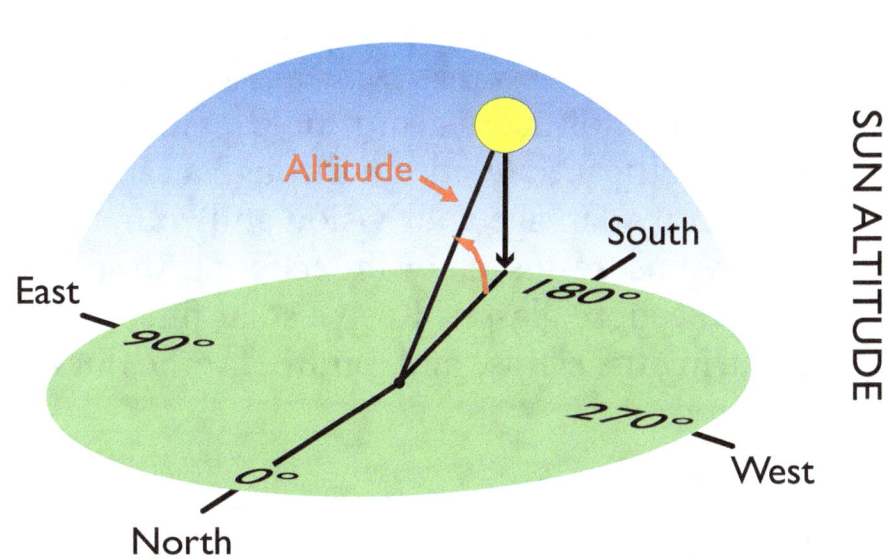

PHOTOGRAPHY

SUN ALTITUDE

Positions given in this book coincide with the time of eclipse totality unless otherwise noted.

Eclipse Data for Select Pennsylvania Locations

LOCATION	TOTALITY START (EDT)
CORRY	3:17:15PM
ERIE	3:16:20PM
GREENVILLE	3:16:26PM
MEADVILLE	3:16:31PM
NORTH EAST	3:16:46PM
YOUNGSVILLE	3:18:11PM

Eclipse Photography

Photographing an eclipse is an exciting challenge, as the moon's shadow moves near 1,600MPH. There is an element of danger and the pressure of time. Looking at the unfiltered sun through a camera can permanently damage your vision and your camera. If you are unsure, just enjoy the eclipse with specially designed eclipse glasses. Keep a solar filter on your lens during the eclipse and remove for the duration of totality!

Partial Vs. Total Solar Eclipse

To successfully and safely photograph a partial and total eclipse, it is important to understand the difference between the two. A solar eclipse occurs when the moon is positioned between the sun and Earth. The region where the shadow of the moon falls upon Earth's surface is where a solar eclipse is visible.

The moon's shadow has two parts—the penumbral shadow and the umbral shadow. The penumbral shadow is the moon's outer shadow where partial solar eclipses can be observed. Total solar eclipses can only be seen within the umbral shadow, the moon's inner shadow.

You cannot say you've seen a total eclipse when all you saw was a partial solar eclipse. It is like saying you've watched a concert, but in reality, you only listened outside the arena. In both cases, you have missed the drama and the action.

PHOTOGRAPHING THE TOTAL SOLAR ECLIPSE

Aside from the region where the outer shadow of the moon is cast, a partial solar eclipse is also visible before a total solar eclipse within the inner shadow region. In both cases, it is imperative to use a solar filter on the lens for both photography and safety reasons. This is the only difference between taking a partial eclipse and a total eclipse photograph of the sun.

To photograph a total solar eclipse, you must be within the Path of Totality, the surface of the Earth within the moon's umbral shadow.

THE CHALLENGE

A total solar eclipse only lasts for a couple of min-

PHOTOGRAPHY

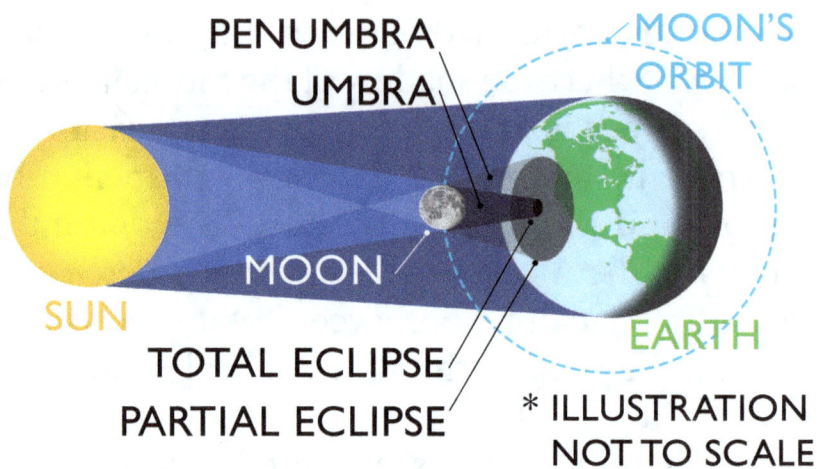

utes. It is brief, but the scenario it brings is unforgettable. Seeing the radiant sun slowly being covered by darkness gives the spectator a high level of anticipation and indescribable excitement. Once the moon completely covers the sun's radiance, the corona is finally visible. In the darkness, the sun's corona shines, capturing the crowd's full attention. Watching this phenomenon is a breathtaking experience.

Amidst all the noise, cheering, and excitement, you have less than a few minutes to take a perfect photograph. The key to this is planning. You need to plan, practice, and perfect what you will do when the big moment arrives because there is no replay. You only have a short time to capture the totality and the sun's corona using different exposures.

PLAN, PRACTICE, PERFECT

It is important to practice photographing before the actual phenomenon arrives. Test your chosen imaging setup for flaws. Rehearse over and over

until your body remembers what you will do from the moment you arrive at your chosen spot to the moment you pack up and leave the area.

You will discover potential problems regarding vibrations and focus that you can address immediately. This minimizes the variables that might affect your photographs at the most critical moment.

It's common for experienced eclipse chasers to lose track of what they plan to do. Write down what you expect to do. Practice it time and again. Play annoying, distracting music while you practice. Try photographing in the worst weather possible. Do anything you can to practice under pressure. Eclipse day is not the time to practice.

Once the sun is completely covered, don't just take photographs. Capture the experience and the image of the total solar eclipse in your mind as well. Set up cameras around you to record not just the total solar eclipse but also the excitement and reaction of the crowd.

PHOTOGRAPHY

ECLIPSE PHOTOGRAPHY GEAR

What do you need to photograph the total eclipse? There are only a few pieces of equipment that you'll need. Preparing to photograph an eclipse successfully takes time. Not only do you have to be skilled and have the right gear, you have to be in the correct place.

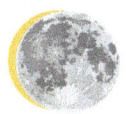

BASIC ECLIPSE PHOTOGRAPHY EQUIPMENT

- Solar viewing glasses (verify authenticity)
- Lens solar filter
- Minimum 300mm lens
- Stable tripod that can be tilted to 60° vertical
- High-resolution DSLR
- Spare batteries for everything
- Secondary camera to photograph people, the horizon, etc.
- Remote cable or wireless release

ADDITIONAL ITEMS

- Video camera
- Video camera tripod
- Quality pair of binoculars
- Solar filters for each binocular lens
- Photo editing software

EQUIPMENT TO PREPARE BEFORE THE BIG DAY

A. Solar viewing glasses

You need a pair of solar viewing glasses as the eclipse approaches.

B. Solar Filter

Partial and total eclipse photography is different from normal photography. Even if only 1% of the sun's surface is visible, it is still approximately 10,000 times brighter than the moon. Before totality, use a solar filter on your lens. Do not look at the sun with

your eyes. It can cause irreparable damage to your retinas.

DO NOT leave your camera pointed at the sun without a solar filter attached. The sun will melt the inside of your camera. Think of a magnifying glass used to torch ants and multiply that by one hundred.

C. Lens

To capture the corona's majesty, you need to use a telescope or a telephoto lens. The best focal length, which will give you a large image of the sun's disk, is 400mm and above. You don't want to waste all your efforts by bringing home a small dot where the black disk and majestic corona are supposed to be.

D. Tripod

Bring a stable enough tripod to support your camera properly to avoid unsteady shots and repeated adjustments. Either will ruin your photos. It also needs to be portable in case you need to change locations for a better shot. *Shut off camera stabilization on a tripod!*

E. Camera

You need to remember to set your camera to its highest resolution to capture all the details. Set your camera to:

- 14-bit RAW is ideal, otherwise
- JPG, Fine compression, Maximum resolution

PHOTOGRAPHY

Bracket your exposures. Shoot at various shutter speeds to capture different brightnesses in the corona. Note that stopping your lens all the way down may not result in the sharpest images.

Choose the lowest possible ISO for the best quality while maintaining a high shutter speed to prevent blurred shots. Set your camera to manual. Do not use AUTO ISO. Your camera will be fooled. The night before, test the focus position of your lens using a bright star or the moon.

Constantly double-check your focus. Be paranoid about this. You can deal with a grainy picture. No amount of Photoshop will fix a blurry, out-of-focus picture.

F. Batteries

Remember to bring fresh batteries! Make sure that you have enough power to capture the most important moments. Swap in fresh batteries thirty minutes before totality.

G. Remote release

Use a wired or wireless remote release to fire the camera's shutter. This will reduce the amount of camera vibration.

H. Video Camera

Run a video camera of yourself. Capture all the things you say and do during the totality. You'll be amazed at your reaction.

I. Photo editing software

You will need quality photo editing software to process your eclipse images. Adobe Lightroom and Photoshop are excellent programs to extract the most out of your images. Become well versed in how to use them at least a month before the eclipse.

J. Smartphone applications

The following smartphone applications will aid in your photography planning: Wunderground, Skyview, Photographer's Ephemeris, Sunrise and Sunset Calculator, SunCalc, and Sun Surveyor among others.

CAMERA PHONES

Smartphone cameras are useful for many things but not eclipse photography. An iPhone 6 camera has a 63° horizontal field of view and is 3264 pixels across. If you attempt to photograph the eclipse, the sun will be a measly 30-40 pixels wide depending on the phone. Digital pinch zoom won't help here. If you want *National Geographic* images, you'll need a serious camera and lens, far beyond any smartphone.

Consider instead using a smartphone to run a time-lapse of the entire event. The sun will be minuscule when shot on a smartphone. Think of something else exciting and interesting do to with it. Purchase a Gorilla Pod, inexpensive tripod, or selfie stick and mount the smartphone somewhere unique.

Also, partial and total eclipse light is strange and

ethereal. Consider using that light to take unique pictures of things and people. It's rare and you may have something no one else does.

FOCAL LENGTH & THE SIZE OF SUN

The size of the sun in a photo depends on the lens focal length. A 300mm lens is the recommended minimum on a full-frame (FF) DSLR. Lenses up to this size are relatively inexpensive. For more magnification, use an APS-C (crop) size sensor. Cameras with these sensors provide an advantage by capturing a larger sun.

For the same focal length, an APS-C sensor will

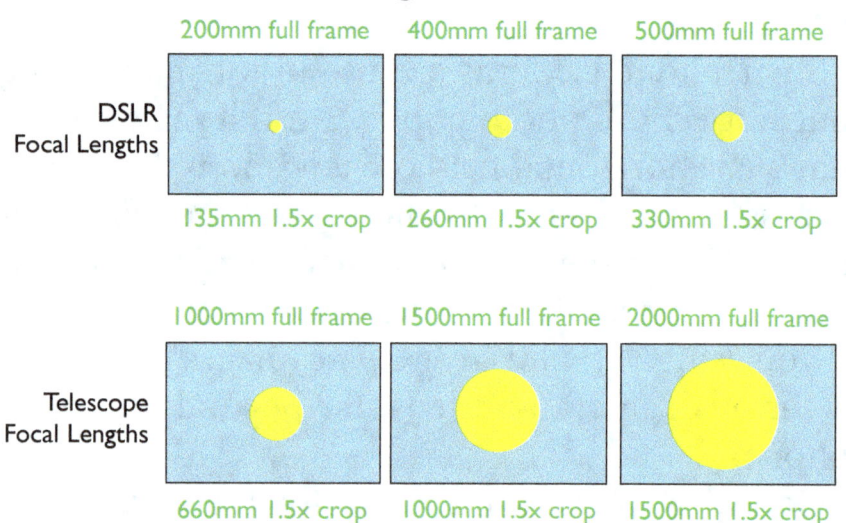

provide a greater apparent magnification of any object. As a consequence, a shorter, less expensive lens can be used to capture the same size sun.

The above figure shows the size of the sun on a camera sensor at various focal lengths. As can be seen with the 200mm lens, the sun is quite small. On a full-frame camera at 200mm, the sun will be 371 pixels wide on a Nikon D810, a 36-megapixel body. A lower resolution FF camera will result in an even smaller sun.

Printing a 24-inch image shot on a Nikon D810 with a 200mm lens at a standard 300 pixels per inch results in a small sun. On this size paper, the sun will

FOCAL LENGTH	FOV FULL FRAME	FF VERT. ANGLE	% OF FF	SUN PIXEL SIZE
14	104° X 81°	81°	0.7%	32.1
20	84° X 62°	62°	0.9%	41.9
28	65° X 46°	46°	1.2%	56.5
35	54° X 38°	38°	1.4%	68.5
50	40° X 27°	27°	2.0%	96.4
105	19° X 13°	13°	4.1%	200.2
200	10° X 7°	7°	7.6%	371.9
400	5° X 3.4°	3.4°	15.6%	765.6
500	4° X 2.7°	2.7°	19.6%	964.2
1000	2° X 1.3°	1.3°	40.8%	2002.5
1500	1.4° X 0.9°	0.9°	58.9%	2892.6
2000	1° X 0.68°	0.68°	77.9%	3828.4

Chart 1: Full-frame camera field of view. The 3rd column is the vertical field of view in degrees. Column 4 is the percentage of the total sensor height that the sun covers. Column 5 is how many pixels wide the sun will be on a 36MP Nikon D810. (Values are estimates)

PHOTOGRAPHY

be a miserly 1.25 inches wide!

Photographing the eclipse with a lens shorter than 300mm will leave you with little to work with. Using a 400mm lens and printing a 24-inch print will result in a 2.5-inch-wide sun. For as massive as the sun is, it is a challenge to take a large photograph of the sun. The sun will appear to move fast with a

PHOTOGRAPHY

FOCAL LENGTH	FOV CROP	CROP VERT DEG	% OF CROP	SUN PIXEL SIZE
14	80° X 58°	58°	0.9%	33.9
20	61° X 43°	43°	1.2%	45.8
28	45° X 31°	31°	1.7%	63.5
35	37° X 25°	25°	2.1%	78.7
50	26° X 18°	18°	2.9%	109.3
105	13° X 8°	8°	6.6%	245.9
200	6.7° X 4.5°	4.5°	11.8%	437.2
400	3.4° X 2°	2°	26.5%	983.7
500	2.7° X 1.8	1.8°	29.4%	1093.0
1000	1.3° X 0.9°	0.9°	58.9%	2186.0
1500	0.9° X 0.6°	0.6°	88.3%	3278.9
2000	0.6° X 0.45°	0.5°	117.8%	4371.9

Chart 2: APS-C Crop sensor camera field of view. The 3rd column is the vertical field of view in degrees. Column 4 is the percentage of the total sensor height that the sun covers. Column 5 is how many pixels wide the sun will be on a 12mp Nikon D300s. (Values are estimates)

500mm lens, too. Plan to adjust.

The big challenge is the cost of the lens. Lenses longer than 300mm are expensive. They also require heavier tripods and specialized tripod heads. The 70-300mm lenses from Nikon, Canon, Tamron, and others are relatively affordable options. It is worth spending time at a local camera shop to try different lenses. Long focal-length lenses are a significant investment, especially for a single event.

To achieve a large eclipse image, you will need a long focal-length lens, ideally at least 400mm. A standard 70-300mm lens set to 300mm will show a small sun. At 500mm, the sun image becomes larger and covers more of the sensor area. The corona will take up a significant portion of the frame. By 1000mm, the corona will exceed the capture area on a full-frame sensor. See the picture in this chapter for sun size simulations for different focal lengths.

SUGGESTED EXPOSURES

To photograph the partial eclipse, the camera must have a solar filter attached. If not, the intense light from the sun may damage (fry) the inside of your camera. This has happened to the author. The exposure depends on the density (darkness) of the solar filter used.

As a starting point, set the camera to ISO 100, f/8, and with the solar filter on, try an exposure of 1/2000. Make adjustments based on the filter used,

PHOTOGRAPHY

histogram, and highlight warning.

Turn on the highlight warning in your camera. This feature is commonly called "blinkies." This warning will help you detect if the image is overexposed or not.

Once the Baily's Beads, prominences, and corona become visible, there will only be a few minutes to take bracketed shots. It will take at least eleven shots to capture the various areas of the sun's corona and stars. The brightness varies considerably. No commercially available camera can capture the incredible dynamic range of the different portions of the delicate corona. This requires taking multiple photographs and digitally combining them afterward.

During totality, try these exposure times at ISO 100 and f/8:
1/4000, 1/2000, 1/1000, 1/250, 1/60, 1/30, 1/15, 1/4, 1/2, 1 sec, and 4 sec.

Disable camera/lens stabilization on a tripod!

PHOTOGRAPHY TIME

Set the camera to full-stop adjustments. It will reduce the time spent fiddling. As an example, the author tried the above shot sequence, adjusting the shutter speed as fast as possible.

It took thirty-three seconds to shoot the above 11 shots using 1/3-stop increments. This was without adjusting composition, focus, or anything else but the shutter speed. When the camera was set to full stop increments, it only took twenty-two seconds to step through the same shutter speed sequence. Use a remote release to reduce camera shake.

Assuming the totality lasts less than two minutes, only four shot sequences could be made using 1/3-stop increments. Yet six shot sequences could be made when the camera was set to full stop steps. Zero time was spent looking at the back LCD to analyze highlights and the histogram.

Now add in the bare minimum time to check the highlight warning. It took sixty-three seconds to shoot and check each image using full stops. And that was without changing the composition to allow for sun movement, bumping the tripod, etc. Looking at the LCD ("chimping") consumed **half** of the totality time.

This test was done in the comfort of home under no pressure. In real world conditions, it may be possible to successfully shoot only one sequence. If you plan to capture the entire dynamic range of the totality, you must practice the sequence until you have it down cold. If you normally fumble with your camera, do not underestimate the difficulty, frustration, and stress of total eclipse photography.

Most importantly, trying to shoot this sequence allowed for zero time to simply look at the totality to enjoy the spectacle.

AVOID LAST MINUTE PURCHASES
You should purchase whatever you think you'll

need to photograph the eclipse early. This event will be nothing short of massive. Remember the hot toy of the year? Multiply that frenzy by a thousand. Everyone will want to try to capture their own photo.

Do not wait until the last few weeks before the eclipse to purchase cameras, lenses, filters, tripods, viewing glasses, and associated material. Consider that the totality of the eclipse will streak from across America. Everyone who wants to photograph the eclipse will order at the same time. If you wait until too late to buy what you need, it's conceivable that solar filters to create a total eclipse photo will be sold out in the United States. All filters sold out during the 2017 total eclipse. Whether this happens or not, do not wait to make your purchases. It may be too late.

Practice

You will need to practice with your equipment. Things may go wrong that you don't anticipate. If you've never photographed a partial or total eclipse, taking quality shots is more difficult than you think. Practice shooting the sequence with a midday sun. This will tell you if you have your exposures and timing correct. Figure out what you need well in advance.

Practice photographing the full moon and stars at night. Capture the moon in full daylight to learn how your camera reacts. Astrophotography is challenging and requires practice.

The August 21, 2017, eclipse as seen in Jackson, WY, shot with a Nikon D800 with an 80-400mm lens set to 340mm. The sun is 644 pixels wide on the 7360x4912 image.

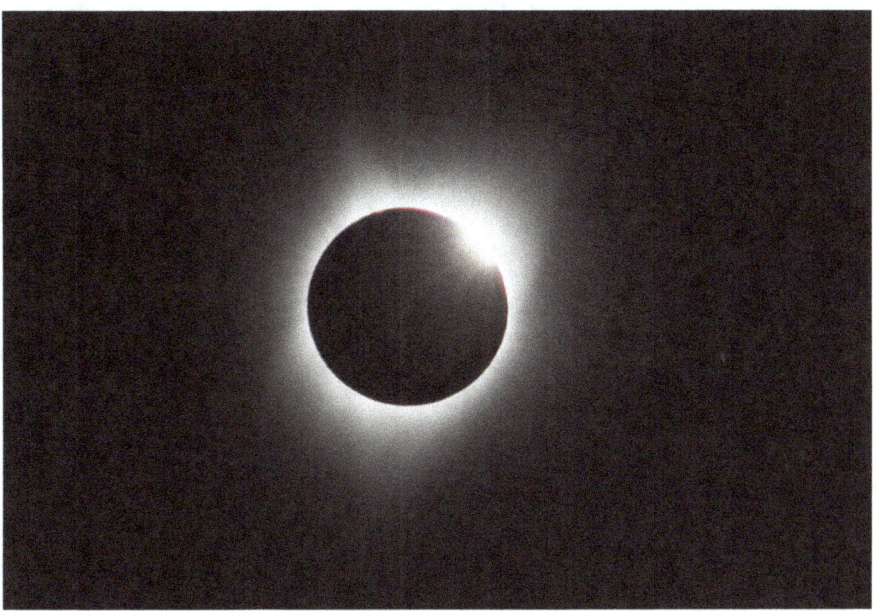

This image is shown straight out of the camera without modification. Even with a high-quality camera and lens, photographing an eclipse is challenging.

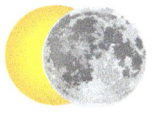

Sun's path from sunrise

Total eclipse position (approximate)

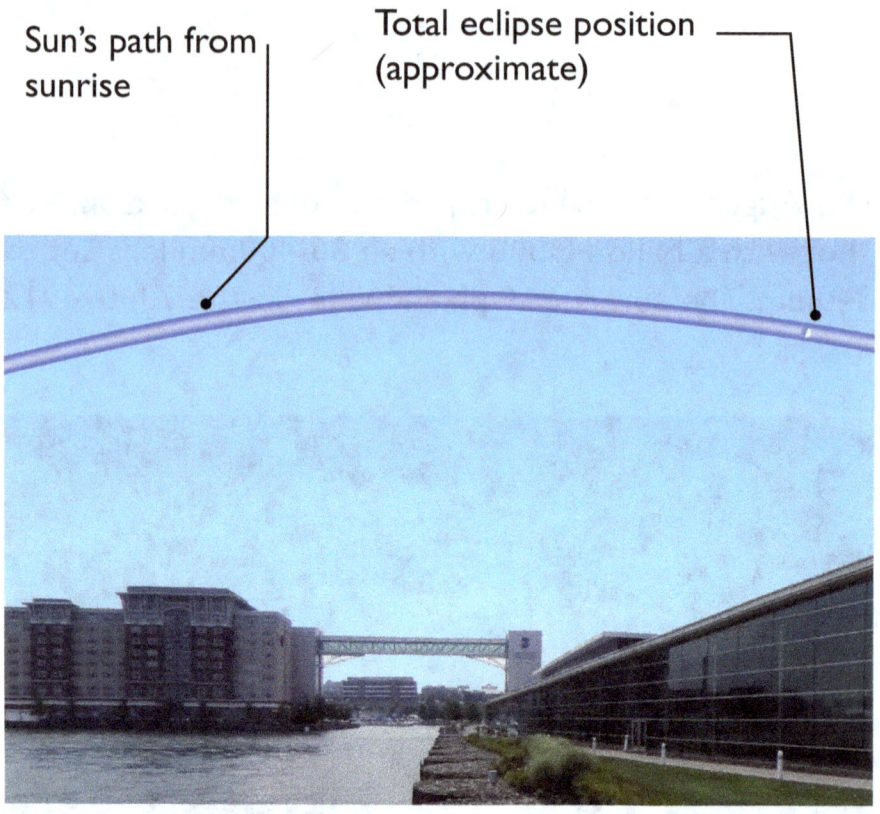

The sun will follow this path on the morning of the eclipse on April 8, 2024.

Image of Erie, Pennsylvania.

Note that this image is a simulation and approximation the sun's path and where the total eclipse may appear from one perspective. Refer to the eclipse position data for a more accurate location.

☉ is the symbol for the sun and first appeared in Europe during the Renaissance.

☾ is the ancient symbol for the moon.

LOCATIONS

Viewing Locations Around Pennsylvania

Tens of thousands of people will travel to and around Pennsylvania to view the total eclipse. There are few obstructions and there is a vast amount of space to view the total eclipse from.

If the weather is questionable, seek out a new location as soon as possible. If you wait until the hour before the eclipse, you may find yourself stuck in traffic.

This section contains popular, alternative, and little-known locations to watch the eclipse. As long as there are no clouds or smoke from fires, the partial eclipse will be viewable from anywhere in the state.

Suggested Total Eclipse View Points

Towns and Cities

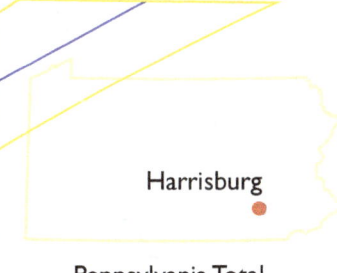

Pennsylvania Total Eclipse Path

- Albion
- Cambridge Springs
- Centerville
- Cochranton
- Corry
- Cranesville
- Erie
- Greenville
- Jamestown
- Meadville
- North East
- Saegertown
- Sheakleyville
- Spartansburg
- Union City
- Wattsburg
- Youngsville

Unique Locations

- Allegheny National Forest
- Erie Bluffs State Park
- Presque Isle State Park
- Pymatuning State Park

LOCATIONS

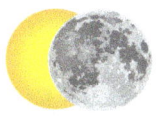

ALBION

Elevation:	890 ft. (296m)
Population:	1,472
Main road/hwy:	US 6N

Albion

OVERVIEW

Albion, a quiet town located near Erie Bluffs, offers visitors a chance to see a juxtaposition of eclipse excitement and small-town serenity. Drive through the historic Harrington Covered Bridge for the town's restaurants and shopping. Albion is a good option if you are looking for a peaceful night's sleep before or after the eclipse.

TOTALITY DURATION

3 minutes 33 seconds

NOTES

Visit Albion's website for updated total eclipse event and lodging information at http://visitpa.com/pa/albion.

EVENT	TIME (EDT)	ALTITUDE	AZIMUTH
SUNRISE	6:51:00AM	0°	79°
ECLIPSE START	2:01:43PM	54°	196°
TOTALITY START	3:15:53PM	47°	224°
TOTALITY END	3:19:27PM	47°	225°
ECLIPSE END	4:30:27PM	36°	243°
SUNSET	7:55:00PM	0°	280°

LOCATIONS

Cambridge Springs

Elevation:	1,155 ft. (385m)
Population:	2,641
Main road/hwy:	US 19

Cambridge Springs

Overview

Located well within the Pennsylvania path of the total eclipse, Cambridge Springs is a pleasant small town to enjoy the total eclipse from. From 1884 to 1915, the town was founded based on the medicinal value of the mineral springs found in the area. The town's triangular park bordered by Church Street, Venago Avenue, and S. Main Street will be one of the best options for viewing the totality with other totality revelers. The park features a World War I memorial that is often surrounded by American flags.

Totality Duration

3 minutes 3 seconds

Notes

https://cambridgespringsborough.com/our-community/

EVENT	TIME (EDT)	ALTITUDE	AZIMUTH
SUNRISE	6:50:00AM	0°	79°
ECLIPSE START	2:02:05PM	54°	197°
TOTALITY START	3:16:31PM	47°	224°
TOTALITY END	3:19:34PM	47°	225°
ECLIPSE END	4:30:46PM	36°	244°
SUNSET	7:54:00PM	0°	281°

LOCATIONS

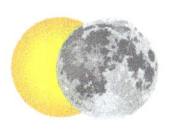

CENTERVILLE

Elevation:	1,306 ft. (435m)
Population:	208
Main road/hwy:	PA 8

Centerville

OVERVIEW

Centerville is a small town with a friendly atmosphere. This town is a hidden getaway nestled in Crawford County and offers 2024 eclipse viewers a chance to slow down and relax. Travelers have several options of unique places to stay from the Caboose Motel to the Bromley's Hillhurst Bed-and-Breakfast. Enjoy home-cooked meals and learn local history in this charming town. Listen to chirping birds as you wander around Centerville. This is a quiet town with various state game lands that will enchant 2024 eclipse seekers.

TOTALITY DURATION

2 minutes 13 seconds

NOTES

https://visitcrawford.org

EVENT	TIME (EDT)	ALTITUDE	AZIMUTH
SUNRISE	6:49:00AM	0°	79°
ECLIPSE START	2:02:27PM	54°	198°
TOTALITY START	3:17:17PM	47°	225°
TOTALITY END	3:19:30PM	46°	226°
ECLIPSE END	4:31:03PM	35°	244°
SUNSET	7:52:00PM	0°	281°

LOCATIONS

COCHRANTON

Elevation:	1,070 ft. (357m)
Population:	1,095
Main road/hwy:	PA 173

Cochranton

OVERVIEW

Cochranton is a small town in Crawford County with overnight and dining opportunities as you rest before or after the eclipse. Wake up to a strong cup of coffee, then head out and about the county for golfing, water sports, hiking, and a day of shopping local at the neighboring town and city downtowns in the county.

TOTALITY DURATION

1 minute 33 second

NOTES

https://visitcrawford.org/plan-your-experience/

EVENT	TIME (EDT)	ALTITUDE	AZIMUTH
SUNRISE	6:50:00AM	0°	80°
ECLIPSE START	2:01:48PM	54°	197°
TOTALITY START	3:17:06PM	47°	225°
TOTALITY END	3:18:39PM	47°	225°
ECLIPSE END	4:30:43PM	36°	244°
SUNSET	7:53:00PM	0°	280°

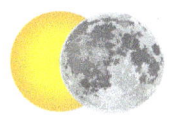

CORRY

Elevation:	1,424 ft. (474m)	Corry
Population:	6,347	
Main road/hwy:	US 6	

OVERVIEW

Corry sits on the path of the 2024 solar eclipse. Book accommodations right in Corry, then for an amazing view of the eclipse, meet up with friends at Mead Park. This 46-acre park offers hiking trails and picnic areas. Enjoy the aroma of grilled meats and yummy sweets as your family and friends gather for the eclipse event. Reserve early to camp overnight in the park.

TOTALITY DURATION

2 minutes 50 seconds

NOTES

https://www.corrychamber.com/listings/mead-park-association/

LOCATIONS

EVENT	TIME (EDT)	ALTITUDE	AZIMUTH
SUNRISE	6:48:00AM	0°	79°
ECLIPSE START	2:02:49PM	54°	198°
TOTALITY START	3:17:15PM	47°	225°
TOTALITY END	3:20:05PM	46°	226°
ECLIPSE END	4:31:13PM	35°	244°
SUNSET	7:52:00PM	0°	281°

CRANESVILLE

Elevation: 945 ft. (315m)
Population: 608
Main road/hwy: PA 18

Cranesville

OVERVIEW

Cranesville is located a few miles south of Erie and will offer 2024 eclipse viewers the chance for a great view of the event. Book your stay at the Four Creeks Bed-and-Breakfast early. You can enjoy this relaxing place to view the eclipse. After the totality event, head northeast to Erie or Westminster for water and outdoor fun. This small town is definitely a getaway option to view the most spectacular sky event over Pennsylvania in recent history.

TOTALITY DURATION

3 minutes 33 seconds

NOTES

https://www.visiterie.com/explore/outdoor-adventures/

EVENT	TIME (EDT)	ALTITUDE	AZIMUTH
SUNRISE	6:51:00AM	0°	80°
ECLIPSE START	2:01:46PM	54°	196°
TOTALITY START	3:15:55PM	47°	224°
TOTALITY END	3:19:29PM	47°	225°
ECLIPSE END	4:30:29PM	36°	243°
SUNSET	7:55:00PM	0°	280°

LOCATIONS

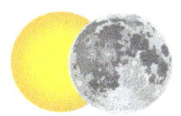

Erie

Elevation:	728 ft. (242m)
Population:	97,369
Main road/hwy:	US 20/I-79

Erie

Overview

Erie, a city of adventures, will delight 2024 solar eclipse sightseers as total darkness covers the city. Adventures abound in Erie for travelers. Visit the numerous downtown shops and restaurants serving American, Asian, and Mediterranean foods. Erie's annual maple syrup festival occurs in early April where visitors enjoy tasty pancakes and fresh sweet maple syrup. Exciting and interesting attractions of this fast-paced city will entertain viewers of the 2024 eclipse.

Totality Duration

3 minutes 42 seconds

Notes

https://www.visitpa.com/region/pennsylvanias-great-lakes-region/erie

LOCATIONS

EVENT	TIME (EDT)	ALTITUDE	AZIMUTH
SUNRISE	6:50:00AM	0°	80°
ECLIPSE START	2:02:23PM	54°	197°
TOTALITY START	3:16:20PM	47°	224°
TOTALITY END	3:20:02PM	46°	225°
ECLIPSE END	4:30:37PM	36°	244°
SUNSET	7:54:00PM	0°	280°

GREENVILLE

Elevation:	410 ft. (125m)
Population:	5,529
Main road/hwy:	PA 358

Greenville

OVERVIEW

Greenville, just north of Pittsburgh, is an ideal stop for 2024 eclipse viewers. While you are waiting for the eclipse, grab a cup of coffee on Main Street, and check out the Canal Museum, the Railroad Museum, and the Greenville Historical Society Waugh House Museum. You can visit the Greenville Museum website for updated event information at https://greenvillemuseumalliance.org/. Book accommodations at the Little Acres Bed-and-Breakfast or one of the nearby hotels.

TOTALITY DURATION

1 minutes 49 seconds

NOTES

https://www.visitmercercountypa.com/about/greenville-pa

EVENT	TIME (EDT)	ALTITUDE	AZIMUTH
SUNRISE	6:52:00AM	0°	79°
ECLIPSE START	2:01:10PM	55°	196°
TOTALITY START	3:16:26PM	47°	224°
TOTALITY END	3:18:15PM	47°	225°
ECLIPSE END	4:30:20PM	36°	244°
SUNSET	7:54:00PM	0°	281°

LOCATIONS

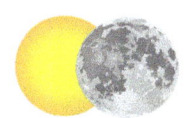

Jamestown

Elevation:	991 ft. (302m)
Population:	582
Main road/hwy:	US 322

Jamestown

Overview

Jamestown is home to Pennsylvania's largest lake, Pymatuning Reservoir. The 2024 solar eclipse will provide viewers with a stunning view at Pymatuning State Park. This park offers visitors the option to see the eclipse from several viewing areas such as the dam, spillway, or lake. Take your kids to the Pymatuning Deer Park then have dinner. Join Jamestown locals and celebrate the 2024 solar eclipse surrounded by nature in this picturesque borough.

Totality Duration

2 minutes 26 seconds

Notes

https://www.dcnr.pa.gov/StateParks/FindAPark/PymatuningStatePark/Pages/default.aspxhttp://city-ofmarionil.gov/

LOCATIONS

EVENT	TIME (EDT)	ALTITUDE	AZIMUTH
SUNRISE	6:52:00AM	0°	80°
ECLIPSE START	2:01:11PM	55°	196°
TOTALITY START	3:16:07PM	47°	224°
TOTALITY END	3:18:33PM	47°	225°
ECLIPSE END	4:30:18PM	36°	244°
SUNSET	7:55:00PM	0°	281°

MEADVILLE

Elevation:	1,400 ft. (466m) Meadville
Population:	13,134
Main road/hwy:	US 322

OVERVIEW

The are several parks to collectively watch the total eclipse from. One of the most picturesque is Diamond Park with its gazebo. If you plan to use a picnic shelter or pavilion in any one of the parks, make sure to send in the rental form as far in advance as possible. Visit the Baldwin-Reynolds House Museum and mansion before the totality to learn more about the city. Originally built by Supreme Court justice Henry Baldwin, this home is one of the best history museums in northwest Pennsylvania.

TOTALITY DURATION

2 minutes 35 seconds

NOTES

https://www.cityofmeadville.org/

EVENT	TIME (EDT)	ALTITUDE	AZIMUTH
SUNRISE	6:51:00AM	0°	79°
ECLIPSE START	2:01:46PM	54°	197°
TOTALITY START	3:16:31PM	47°	224°
TOTALITY END	3:19:07PM	47°	225°
ECLIPSE END	4:30:38PM	36°	244°
SUNSET	7:54:00PM	0°	281°

LOCATIONS

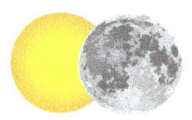

North East

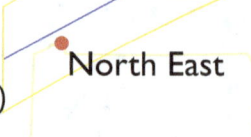

North East

Elevation:	802 ft. (244m)
Population:	4,128
Main road/hwy:	US 20

Overview

North East is a relaxing community of vineyards, wineries, and beautiful views of the southern shore of Lake Erie. The vineyards in the area date back to 1850 when two men, Griffith and Hammond, planted the first vineyards in the region. There are thousands of acres of vineyards and eleven vibrant wineries. Located on the corner of Main Street and North Lake Street, Gibson Park will be a perfect location to see the crescent moon from the total eclipse in from the shadows of the trees and spring leaves.

Totality Duration

3 minutes 40 seconds

Notes

https://nechamber.org/

LOCATIONS

EVENT	TIME (EDT)	ALTITUDE	AZIMUTH
SUNRISE	6:49:00AM	0°	79°
ECLIPSE START	2:02:53PM	54°	198°
TOTALITY START	3:16:46PM	47°	224°
TOTALITY END	3:20:27PM	46°	225°
ECLIPSE END	4:31:06PM	35°	244°
SUNSET	7:53:00PM	0°	281°

SAEGERTOWN

Saegertown

Elevation:	1,112 ft. (370m)
Population:	962
Main road/hwy:	US 19

OVERVIEW

Escape from the crowds and book a room in Saegertown, a small town just ten minutes from the Woodcock Lake Park and campground where you can take in the view from the top of the lake's dam, let your kids loose in the playground, and walk the trail in the Bossard Nature Area. Bring a picnic lunch and watch the eclipse with the whole family.

TOTALITY DURATION

2 minutes 53 seconds

NOTES

http://www.woodcocklakepark.com/index.html

EVENT	TIME (EDT)	ALTITUDE	AZIMUTH
SUNRISE	6:50:00AM	0°	79°
ECLIPSE START	2:01:52PM	54°	197°
TOTALITY START	3:16:26PM	47°	224°
TOTALITY END	3:19:19PM	47°	225°
ECLIPSE END	4:30:40PM	36°	244°
SUNSET	7:54:00PM	0°	281°

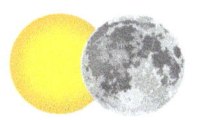

SHEAKLEYVILLE

Elevation:	1447 ft. (482M)	Sheakleyville
Population:	134	
Main road/hwy:	US 19	

OVERVIEW

Sheakleyville, a sleepy town a few miles east of Greenville, is on the path of the 2024 solar eclipse. Camp just ten minutes away at Goddard Park Vacationland Campground along Lake Wilhelm, then spend your day at Maurice K. Goddard Park fishing, hiking, boating, and wildlife spotting. TOTALITY

DURATION

1 minutes 28 seconds

NOTES

https://www.dcnr.pa.gov/StateParks/FindAPark/MauriceKGoddardStatePark/Pages/default.aspx

EVENT	TIME (EDT)	ALTITUDE	AZIMUTH
SUNRISE	6:51:00AM	0°	79°
ECLIPSE START	2:01:29PM	55°	197°
TOTALITY START	3:16:53PM	47°	224°
TOTALITY END	3:18:21PM	47°	225°
ECLIPSE END	4:30:32PM	36°	244°
SUNSET	7:54:00PM	0°	281°

LOCATIONS

Spartansburg

Elevation:	1,447 ft. (482m)
Population:	294
Main road/hwy:	IL 89

Spartansburg

Overview

You'll find this little town about thirty miles southeast of Erie. Visitors can dine on local Pennsylvania Dutch cuisine and shop local artist's creations from handmade quilts to hand-carved furniture. Meet at Clear Lake Park for a quiet picnic or travel on the East Branch Trail alongside Amish travelers. Visitors may want to browse Amish goods for a handcrafted souvenir to take home.

Totality Duration

2 minutes 30 seconds

Notes

https://spartansburg.org

EVENT	TIME (EDT)	ALTITUDE	AZIMUTH
SUNRISE	6:49:00AM	0°	79°
ECLIPSE START	2:02:39PM	54°	198°
TOTALITY START	3:17:18PM	47°	225°
TOTALITY END	3:19:48PM	46°	226°
ECLIPSE END	4:31:09PM	35°	244°
SUNSET	7:52:00PM	0°	282°

LOCATIONS

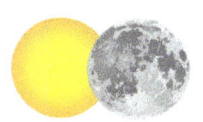

UNION CITY

Elevation:	1,263 ft. (421m)
Population:	3,181
Main road/hwy:	US 6

Union City

OVERVIEW

Union City, historical yet modern, is just south of Erie. Stroll along Main Street, take in the architecture, and learn about Union City's humble beginnings through its collection of early artifacts at the Union City Historical Museum which displays how early settlers lived and how those settlements impacted Pennsylvania. Book overnight accommodations and enjoy the solar eclipse from this historical town.

TOTALITY DURATION

3 minutes 4 seconds

NOTES

http://unioncitypa.us/

LOCATIONS

EVENT	TIME (EDT)	ALTITUDE	AZIMUTH
SUNRISE	6:49:00PM	0°	80°
ECLIPSE START	2:02:30PM	54°	197°
TOTALITY START	3:16:51PM	47°	225°
TOTALITY END	3:19:55PM	46°	225°
ECLIPSE END	4:31:00PM	35°	244°
SUNSET	7:53:00PM	0°	280°

Wattsburg

Elevation:	1,286 ft. (428m)
Population:	385
Main road/hwy:	PA 89

Overview

Wattsburg, a small town in northwest Pennsylvania, offers visitors a unique view of the 2024 solar eclipse. Stay overnight in Wattsburg, then take a short trip north on I-89 toward Lake Erie for a host of activities and events before or after the eclipse. Wattsburg will offer visitors a small-town experience at a reasonable budget.

Totality Duration

3 minutes 17 seconds

Notes

https://www.visiterie.com/

EVENT	TIME (EDT)	ALTITUDE	AZIMUTH
SUNRISE	6:49:00AM	0°	79°
ECLIPSE START	2:02:39PM	54°	198°
TOTALITY START	3:16:50PM	47°	224°
TOTALITY END	3:20:08PM	46°	225°
ECLIPSE END	4:31:03PM	35°	244°
SUNSET	7:53:00PM	0°	281°

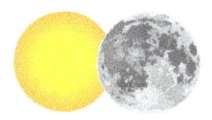

YOUNGSVILLE

Elevation:	1,201 ft. (400m)
Population:	1,630
Main road/hwy:	US 6

OVERVIEW

Youngsville, located near the Allegheny National Forest, sits at the intersection of Route 6 and Route 27 in Warren County. Stroll around downtown, browse the shops, and take home crafted jewelry or local art. Locals proudly call it "the biggest little town on the map." Youngsville's Facebook community page will be a perfect place to find out about total eclipse events at https://www.facebook.com/groups/418847134828749/.

TOTALITY DURATION

1 minute 45 seconds

NOTES

https://www.wcvb.net/

LOCATIONS

EVENT	TIME (EDT)	ALTITUDE	AZIMUTH
SUNRISE	6:47:00AM	0°	79°
ECLIPSE START	2:03:14PM	54°	199°
TOTALITY START	3:18:11PM	46°	226°
TOTALITY END	3:19:56PM	46°	226°
ECLIPSE END	4:31:32PM	35°	245°
SUNSET	7:51:00PM	0°	281°

ALLEGHENY NATIONAL FOREST

Elevation:	Multiple
Main road/hwy:	Multiple

Allegheny National Forest

OVERVIEW

Allegheny National Forest will provide 2024 solar eclipse viewers with an event tailored to individual interests. From backpackers to campers, the forest will delight every traveler. Visitors can choose to camp on the grounds in tents or at the Farnsworth Cabin. Another interesting attraction for 2024 eclipse viewers is the Longhouse National Scenic Byway. This thirty-six-mile stretch of road cuts through the Allegheny National Forest and provides breathtaking views.

TOTALITY DURATION

Various depending on location.

NOTES

https://visitanf.com/

Times are for the Allegheny Reservoir

EVENT	TIME (EDT)	ALTITUDE	AZIMUTH
SUNRISE	6:45:00AM	0°	79°
ECLIPSE START	2:03:55PM	54°	199°
TOTALITY START	3:18:50PM	46°	226°
TOTALITY END	3:20:25PM	46°	227°
ECLIPSE END	4:31:57PM	46°	227°
SUNSET	7:49:00PM	0°	281°

LOCATIONS

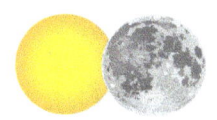

ERIE BLUFFS STATE PARK

Elevation: 708 ft. (236m)
Main road/hwy: PA 5

Erie Bluffs
State Park

OVERVIEW

Erie Bluffs State Park is located twelve miles west of Erie and will be a majestic backdrop for the 2024 solar eclipse. Take a picnic lunch with you to the park, then once the park closes at dusk, venture to the shoreline of Lake Erie for shopping and dining at the area's restaurants. There are several opportunities for lodging, but be sure to book early to beat the crowds.

TOTALITY DURATION

3 minutes 43 seconds

NOTES

https://www.dcnr.pa.gov/StateParks/FindAPark/ErieBluffsStatePark/Pages/default.aspx

EVENT	TIME (EDT)	ALTITUDE	AZIMUTH
SUNRISE	6:51:00AM	0°	79°
ECLIPSE START	2:01:50PM	54°	196°
TOTALITY START	3:15:52PM	47°	223°
TOTALITY END	3:19:36PM	47°	225°
ECLIPSE END	4:30:28PM	36°	243°
SUNSET	7:55:00PM	0°	281°

LOCATIONS

Presque Isle State Park

Elevation:	580 ft. (193m)
Main road/hwy:	PA 832

Presque Isle
State Park

Overview

Presque Isle State Park is located a fifteen-minute ride north of Erie. This peninsula will provide viewers with a unique experience. Visitors can gaze across the island at the lighthouse and enjoy amazing views of the 2024 solar eclipse. Travelers will enjoy the sandy lake shore and its feathery occupants before venturing inland to downtown Erie. Join the excitement and adventure at Presque Isle State Park in 2024 for the solar eclipse.

Totality Duration

3 minutes 44 seconds

Notes

https://www.dcnr.pa.gov/StateParks/FindAPark/PresqueIsleStatePark/Pages/default.aspx

EVENT	TIME (EDT)	ALTITUDE	AZIMUTH
SUNRISE	6:50:00AM	0°	79°
ECLIPSE START	2:02:24PM	54°	197°
TOTALITY START	3:16:19PM	47°	224°
TOTALITY END	3:20:03PM	46°	225°
ECLIPSE END	4:30:47PM	36°	244°
SUNSET	7:54:00PM	0°	281°

LOCATIONS

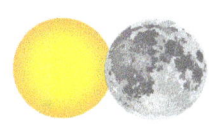

PYMATUNING STATE PARK

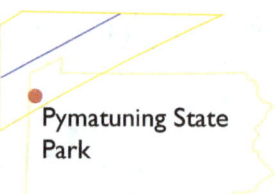

Pymatuning State Park

| Elevation: | Multiple |
| Main road/hwy: | Various |

OVERVIEW

Located on the border of Pennsylvania and Ohio, Pymatuning State Park is home to the crescent-shaped 17,088-acre Pymatuning Reservoir. This park is split between the two states and is 16,892 acres. The reservoir supports fishing, boating, wildlife watching, camping, and outdoor adventuring. Fish are reported to be plentiful at the reservoir's spillway. Bald eagles are often sighted along the lakeshore, thrilling visitors.

TOTALITY DURATION

Various depending on location.

NOTES

https://www.dcnr.pa.gov/StateParks/Find-APark/PymatuningStatePark/Pages/default.aspxreserveamerica.com/explore/wayne-fitzgerrell-state-recreation-area/IL/454481/overview

EVENT	TIME (EDT)	ALTITUDE	AZIMUTH
SUNRISE	6:52:00AM	0°	79°
ECLIPSE START	2:01:23PM	54°	196°
TOTALITY START	3:15:55PM	47°	224°
TOTALITY END	3:19:00PM	47°	225°
ECLIPSE END	4:30:21PM	36°	244°
SUNSET	7:55:00PM	0°	281°

LOCATIONS

Remember the Pennsylvania Total Eclipse
April 8, 2024

Who was I with? _____

_____

_____

_____

_____

What did I see? _____

_____

_____

_____

_____

_____

_____

_____

What did I feel? _____

_____

_____

_____

_____

_____

_____

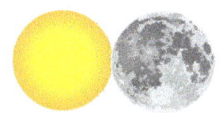

_____

_____

_____

_____

_____

What did the people with me think? _____

_____

_____

_____

_____

_____

_____

_____

_____

Where did I stay?_____

_____

_____

_____

_____

_____

_____

_____

Aaron Linsdau, Polar Explorer & Motivational Speaker

Aaron Linsdau is a polar explorer and motivational speaker. He energizes audiences with life and business lessons that stick. He teaches how to build resilience to overcome adversity by attitude maintenance

He holds the world record for the longest expedition in days from Hercules Inlet to the South Pole. Aaron is the second only American to complete the trip alone, eating seventy pounds of butter on the expedition.

Aaron collaborates with organizations to deliver the right message for the audience. He connects his stories to business realities. Aaron loves inspiring audiences. Book Aaron for your next event today.

"Never Give Up"

ADVERSITY • ATTITUDE • RESILIENCE • RISK • SAFETY

www.ncexped.com

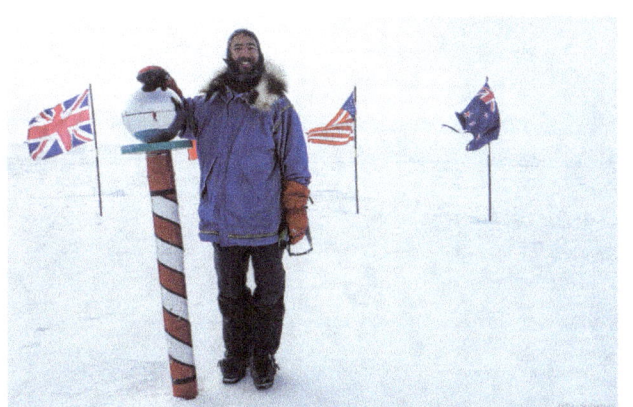

Smartphone link

Aaron at the South Pole after 82 days alone in Antarctica.

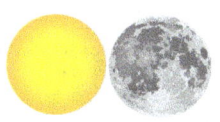

www.ingramcontent.com/pod-product-compliance
Lightning Source LLC
Chambersburg PA
CBHW071225220526
45468CB00002B/742